SPACE EXPLORATION
TRIUMPHS AND TRAGEDIES

Sonya Newland

CRABTREE
Publishing Company
www.crabtreebooks.com

Crabtree Publishing Company
www.crabtreebooks.com

Author: Sonya Newland
**Publishing plan research
and development**: Reagan Miller
Project coordinator: Crystal Sikkens
Editors: Sonya Newland, Crystal Sikkens
Proofreader: Janine Deschenes
Designer: Tim Mayer (Mayer Media)
Picture researcher: Sonya Newland
Cover design: Ken Wright
**Production coordinator and
prepress technician**: Ken Wright
Print coordinator: Margaret Amy Salter
Production coordinated by:
White-Thomson Publishing

Photographs:
Alamy: Heritage Image Partnership: p. 16;
RIA Novosti: p. 17; Getty Images: Sovfoto:
p. 13; NASA: pp. 4, 5, 10, 14, 18, 19, 20, 21,
22–23, 24, 25, 26, 27, 28, 30–31, 31, 32, 34, 35,
37, 39; JPL-Caltech/STScI/CXC/SAO: pp.
1, 36–37; Pat Rawlings: pp. 3, 40–41; SDO:
p. 8; NSSDC: p. 12; Lynda Brammer: p. 33;
/JPL-Caltech/Univ. of Arizona: pp. 38–39;
JPL/DLR: p. 43; Shutterstock: cover; Filip
Fuxa: pp. 6–7; Jorg Hackemann: p. 11; Digital
Storm: pp. 42–43; Wikimedia: p. 9; Laika ac:
p. 15; Virgin Galactic/Mark Greenberg: pp.
44–45.

Library and Archives Canada Cataloguing in Publication

Newland, Sonya, author
        Space exploration : triumphs and tragedies / Sonya Newland.

(Crabtree chrome)
Includes index.
Issued in print and electronic formats.
ISBN 978-0-7787-2292-2 (bound).--ISBN 978-0-7787-2231-1
(paperback).--ISBN 978-1-4271-8088-9 (html)

        1. Outer space--Exploration--Juvenile literature.
2. Astronautics--Juvenile literature. I. Title. II. Series:
Crabtree chrome

TL793.N48 2016   j629.4   C2015-907949-7
                                    C2015-907950-0

Library of Congress Cataloging-in-Publication Data

CIP available at Library of Congress

Crabtree Publishing Company
www.crabtreebooks.com   1-800-387-7650

Printed in Canada/022016/MA20151130

Published in Canada
Crabtree Publishing
616 Welland Ave.
St. Catharines, ON
L2M 5V6

Published in the United States
Crabtree Publishing
PMB 59051
350 Fifth Avenue, 59th Floor
New York, New York 10118

Published in the United Kingdom
Crabtree Publishing
Maritime House
Basin Road North, Hove
BN41 1WR

Published in Australia
Crabtree Publishing
3 Charles Street
Coburg North
VIC 3058

Contents

Disaster in Space

"Houston, we've had a problem here." The voice of astronaut Jack Swigert crackled through to NASA's Mission Control in Texas. **CapCom** Jack Lousma asked Swigert to repeat the message. His fellow astronaut, Commander Jim Lovell, took the radio. "We've had a main B bus undervolt," he reported. Something had gone terribly wrong on board *Apollo 13*.

▲ *Mission Control talks to astronaut Fred Haise (on the screen) during the* Apollo 13 *mission.*

Bring Them Home

Two days into their journey to the Moon, the astronauts were more than 200,000 miles (322,000 kilometers) from Earth. They were running out of air and they no longer had enough electricity to get them home. It was Mission Control's job to bring them back alive.

◀ *NASA's motto is "For the Benefit of All."*

NASA is the USA's National Aeronautics and Space Administration. It was set up in 1958 to develop technology that would help people learn more about—and travel in—space.

CapCom: the person on Earth in charge of talking to astronauts

Staring into Space

For hundreds of years, humans have looked up at the night sky and asked questions about what they saw. What is the Sun? How many other planets are there? How far away are the stars? Ancient **astronomers** did not have telescopes or spacecraft to find the answers. Their knowledge of space was based on what they could see with the naked eye.

▼ *The prehistoric monument Stonehenge, in the U.K., may be linked to astronomy. From a certain position, the Heel Stone (pictured here, in the center) lines up with the sunrise in midsummer, and sunset in midwinter.*

Ancient Astronomy

In ancient times, people set their calendars by the movements of the Sun and Moon. They discovered that objects in space had an effect on the weather and the tides. Tides are the way the sea rises and falls throughout the day. They saw their heroes and gods in the patterns of the stars called constellations.

The science of astronomy, or the study of space, may have begun in China around 6,000 years ago. The name comes from two Greek words, meaning "star" and "law."

astronomers: scientists who study objects in space

Astronomy in Ancient Greece

ancient Greeks began applying math to the mo
bjects in space. They discovered that Earth move
and the Sun and that the Moon traveled around
ek astronomers were the first to realize that Earth
ound. They also worked out what caused **eclipse**

he Greek astronomer
agoras figured out that a
eclipse was caused when
Moon passed between the
and Earth.

"Everything has a natural explanation.
The Moon is not a god, but a great rock,
and the Sun a hot rock."

Ancient Greek astronomer, Anaxagoras

A Closer Look

In 1609, the Italian astronomer Galileo Galilei was the first person to look at the night sky through a telescope. He made many discoveries, including the rings around Saturn and Jupiter's moons. Telescopes showed that there was a lot more to space than anyone had ever imagined. But it was many years later before people had the technology to travel into space.

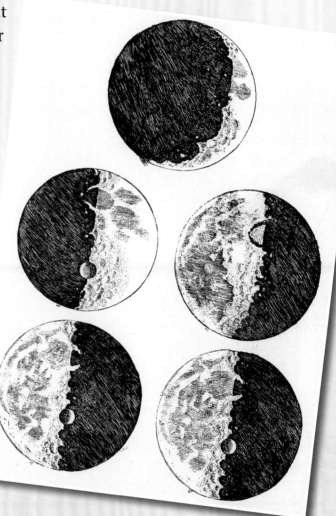

▲ *Galileo drew sketches of the Moon, showing the mountains and craters he saw on its surface.*

eclipses: when an object in space blocks the light from another object

The Space Race

The Superpowers

After World War II, the United States and the Soviet Union (USSR, or Russia) were the most powerful nations on Earth. They each wanted to be the first to send people to space. Both countries put their best scientists to work figuring out how to build machines that could travel beyond Earth.

◀ *German scientist Wernher von Braun helped the Americans develop their rocket technology after the war.*

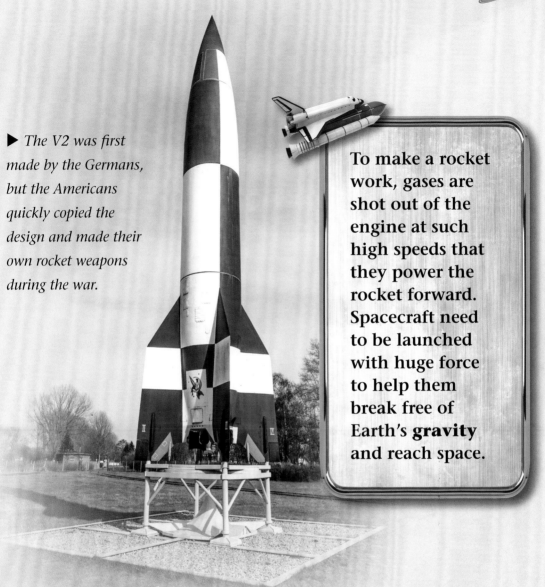

▶ *The V2 was first made by the Germans, but the Americans quickly copied the design and made their own rocket weapons during the war.*

To make a rocket work, gases are shot out of the engine at such high speeds that they power the rocket forward. Spacecraft need to be launched with huge force to help them break free of Earth's **gravity** and reach space.

Rocket Technology

During the war, scientists had developed rockets to use as weapons. These deadly devices could travel great distances. During a test in 1942, a German V2 rocket had shot to 60 miles (100 kilometers) above Earth's surface—the very edge of space. Afterward, experts realized they could use rockets like this to launch spacecraft.

gravity: the force of attraction between two objects

▶ Sputnik 1 *sent out radio signals that gave scientists a lot of new information about Earth's atmosphere.*

Dawn of the Space Age

In 1957, the USSR launched the first **satellite**, which they called *Sputnik 1*. It spent three months in space —a journey of more than 43 million miles (69 million kilometers). This small metal ball was the first human-made object to circle Earth in space. The event marked the start of a period of time known as the "Space Age."

Sputnik 1 launched our way of life today. There are now around 2,500 human-made satellites moving around Earth. They allow us to use devices such as cells phones and GPS systems. They also provide weather information.

First Moon Landing

Two years later, the Soviets also became the first to land a craft on the Moon. *Luna 2* was a small spacecraft, only 3 feet (1 meter) across, designed to crash-land on the lunar (Moon's) surface. *Luna 2* was destroyed on impact, but it proved that a spacecraft could reach the Moon.

▲ Luna 2 *crashed into the Moon at a speed of about 7,200 mph (11,600 km/h).*

satellite: a device that collects or sends information

Space Monkeys

Before the Soviets launched *Sputnik 1*, the Americans had achieved their own "first" by sending a monkey, named Albert II, into space on a V2 rocket. Albert II soared to 83 miles (134 kilometers) above Earth's surface, but he did not survive the return journey. Experiments like this gave scientists an idea of how living creatures coped with being in space.

▲ *A rhesus monkey named Sam flew to 63 miles (101 kilometers) above Earth in 1959. He survived the journey.*

Laika

The Soviets were also experimenting with animals in space. Their second satellite, *Sputnik 2*, carried a passenger—a dog called Laika. She was the first animal to **orbit** Earth. Scientists did not yet know how to bring a spacecraft back once it was in orbit, so Laika died in space.

A satellite is kept in orbit by the pull of gravity from Earth. You can think of it like swinging a ball in a circle on a piece of string. The string is like gravity—without it, the satellite would shoot off into space.

◀ *There is a monument to Laika in Moscow, Russia.*

orbit: to move around a circular object

"Nothing will stop us. The road to the stars is steep and dangerous. But we are not afraid...Space flights can't be stopped."

Russian cosmonaut, Yuri Gagarin

▲ *Fighter pilot Yuri Gagarin's space flight took 108 minutes.*

More Soviet Successes

On April 12, 1961, Russian **cosmonaut** Yuri Gagarin made history—he became the first man in space. He made one orbit of Earth in *Vostok 1*, traveling at an incredible speed of 17,500 mph (28,000 km/h)! Less than two hours after launch, Gagarin parachuted out of the spacecraft and drifted safely back to Earth.

Female First

Russian Valentina Tereshkova was the first woman in space. On June 16, 1963, she took off in a *Vostok 6* spacecraft from the world's first space-launching facility, the Baikonur Cosmodrome. Tereshkova spent nearly three days in orbit, traveling over one million miles (1.6 million km). Today, more than 50 years later, Tereshkova wants to go back into space. She has said she would like to join a mission to Mars!

▲ *Tereshkova worked in a factory before she was chosen to be a cosmonaut. After her spaceflight she became a hero to the Russian people.*

cosmonaut: a Russian astronaut

Project Apollo

Launching Apollo

On May 25, 1961, President Kennedy made a big announcement. He said that the USA would land a man on the Moon "before this decade is out." Kennedy was right. In July 1969, astronauts Neil Armstrong and Buzz Aldrin became the first humans to set foot on another world. But it was a long journey, in more ways than one.

▲ *This photograph shows* Apollo 1 *astronauts in training.*

Tragedy Strikes

In 1961, NASA announced its plan to begin a series of manned and unmanned missions to study the Moon. Unfortunately, this series of missions, called Project Apollo, got off to a tragic start. On January 17, 1967, during a practice mission, a fire broke out in the small **command module** of *Apollo 1*. The fire killed all three astronauts inside.

▲ *The* Apollo 1 *astronauts: Gus Grissom, Edward White, and Roger Chaffee.*

"We are going to have failures. There are going to be sacrifices made in the program. ...The [discovery] of space is worth the risk of life."

Virgil "Gus" Grissom, one of the three astronauts who died in *Apollo 1*

command module: the control center in a spacecraft

Videos from Space

After the terrible fate of *Apollo 1*, everyone was nervous about the next manned mission. But all went well with *Apollo 7*. This mission's goal was to orbit Earth and test the program's new command module. The astronauts also took a video camera with them. For the first time, people on Earth saw live pictures of astronauts floating in their spacecraft.

▲ The Apollo 7 *astronauts (Don Eisele, Walter Schirra, and Walter Cunningham) are shown here by the main hatch of the command module.*

The Far Side

The next step in the program was to take the command module on an orbit of the Moon. In December 1968, *Apollo 8* carried astronauts Jim Lovell, Bill Anders, and Frank Borman deeper into space than humans had ever been. *Apollo 8* was the first time anyone had ever seen the far side of the Moon!

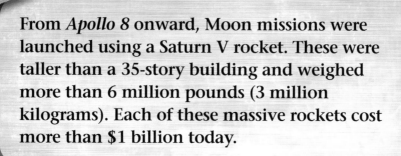

From *Apollo 8* onward, Moon missions were launched using a Saturn V rocket. These were taller than a 35-story building and weighed more than 6 million pounds (3 million kilograms). Each of these massive rockets cost more than $1 billion today.

▶ *The* Apollo 8 *astronauts were amazed as they watched earth rise over the lunar **horizon**.*

horizon: the line where land and sky seem to meet

The Eagle Has Landed

July 20, 1969. After a three-day journey from Earth, the *Apollo 11* lunar module *Eagle* touched down on a part of the Moon known as the Sea of Tranquility. Neil Armstrong climbed carefully down the ladder and stepped out onto the surface of the Moon. Buzz Aldrin followed him. In the command module, Michael Collins listened to the **transmissions** between his fellow astronauts and the excited team at Mission Control back on Earth.

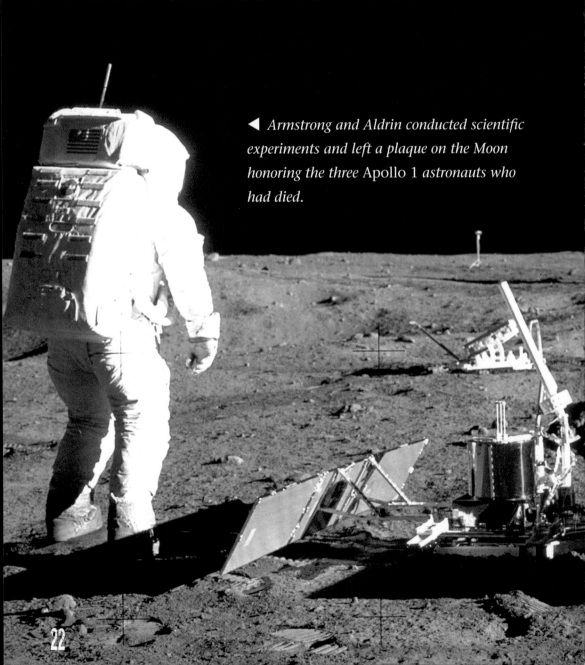

◀ *Armstrong and Aldrin conducted scientific experiments and left a plaque on the Moon honoring the three* Apollo 1 *astronauts who had died.*

One Small Step

Once on the surface, Armstrong and Aldrin took a telephone call from President Nixon. Then they set about collecting samples of Moon rocks and dust. They also took plenty of photographs. After nearly a day on the surface, they climbed back into the *Eagle* and headed back to the command module. History had been made.

The lunar module was one of three parts that made up the *Apollo 11* craft. There was also the service module and the command module, called *Columbia*. The command module was where the astronauts lived during their journey. It was the only part that returned to Earth.

transmissions: radio messages

Apollo 13

Apollo 13 was heading to the Moon to explore an area called the Fra Mauro Highlands. Two days into its journey, there was an explosion in one of the oxygen tanks, leaving the astronauts short on air. The tanks also ran the fuel cells that provided the electricity. Now, they wouldn't have enough power for the return journey. Experts on the ground desperately tried to find a way to fix the damaged spacecraft. Then they had to get *Apollo 13* home.

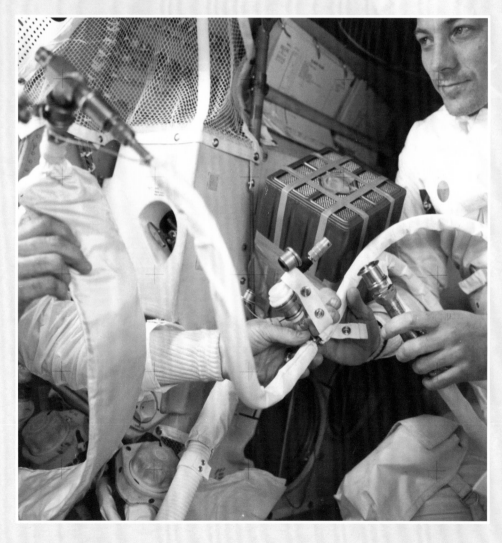

▲ *Mission Control helped the astronauts fix the spacecraft using only the equipment they had on board.*

▲ *The* Apollo 13 *astronauts are lifted to safety from the* Odyssey *command module after their splashdown.*

A Successful Failure

The astronauts guided the spacecraft around the Moon. They used its gravity to boost them homeward. Then they switched off everything they could to save as much power as possible. They survived for days in near-freezing conditions, but at last they made a safe **splashdown** in the Pacific Ocean.

NASA described *Apollo 13* as a "successful failure." The astronauts did not reach the Moon, but they returned safely to Earth against all the odds.

splashdown: when a spacecraft lands in the sea

Roving the Moon

Apollo 15 was the ninth manned mission to the Moon.
This was the first time that the special vehicle called the
Lunar Roving Vehicle was used to explore the landscape
of the Moon. The astronauts also carried out several **EVAs**
(extra-vehicular activity) both from the surface of the
Moon and in space.

▼ *Powered by electricity, the Lunar Roving Vehicle had a top speed of
8 mph (13 km/h).*

"No single space project in this period
will be more impressive to mankind,
or more important for the long-range
exploration of space."

U.S. President John F. Kennedy, talking
about the *Apollo* missions

▼ Apollo 17 *was the first and only Moon mission to be launched at night.*

Last Men on the Moon

In December 1972, the *Apollo 17* astronauts became the last men to stand on the Moon. One of them, Harrison Schmitt, was the first professional scientist to go there. Nearly 50 years later, NASA is once again planning a Moon mission. By 2020, humans may once again walk on the lunar surface.

EVA: when astronauts move around in space outside their spacecraft

What's Next?

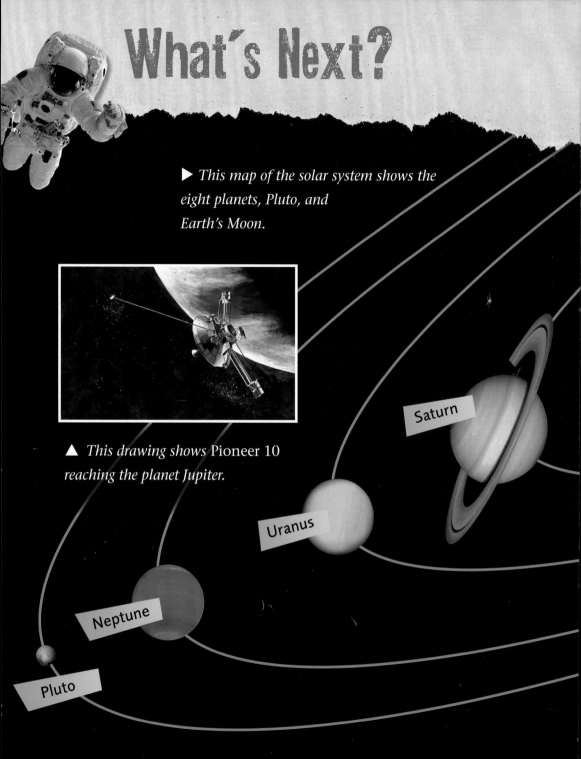

▶ *This map of the solar system shows the eight planets, Pluto, and Earth's Moon.*

▲ *This drawing shows* Pioneer 10 *reaching the planet Jupiter.*

Saturn

Uranus

Neptune

Pluto

Pioneering Probes

While the *Apollo* missions were going on, NASA was also exploring other parts of the solar system. It used **probes** to find out more about the other planets, including the *Pioneer* probes. *Pioneer* probes were sent to Jupiter, Saturn, and Venus. They were even sent to the Sun.

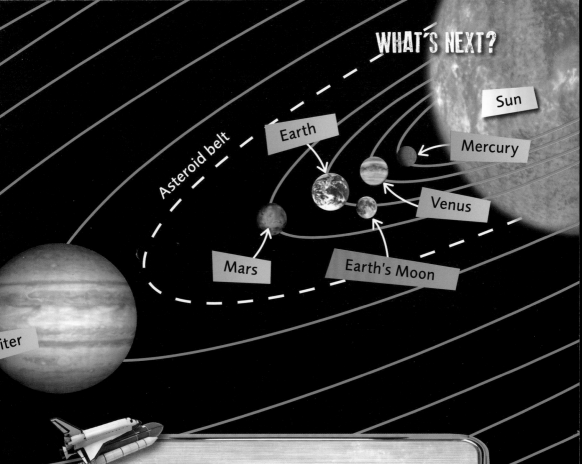

Sun

Earth

Mercury

Asteroid belt

Venus

Mars

Earth's Moon

iter

Pioneers 10 and *11* each carry a golden plaque. These show pictures of a man and a woman. They also explain where the space probes came from, in case they are found by other life-forms.

The Edges of the Solar System

Pioneers 10 and *11* were designed to travel farther than any other spacecraft had before. They journeyed to the part of space called the asteroid belt, which is an area between Mars and Jupiter, then on to the outer solar system. Scientists learned a lot from the information the *Pioneers* sent back, before they lost contact with *Pioneer 11* in 1995 and *Pioneer 10* in 2003.

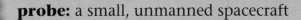

probe: a small, unmanned spacecraft

▶ Voyager 1 *took this photograph of the Great Red Spot, a huge storm on Jupiter that is three times the size of Earth!*

The *Voyager* Probes

The *Voyager* probes were launched in 1977. Their main mission was to send back information about the outer planets—Jupiter, Saturn, Uranus, and Neptune. The *Voyagers* showed scientists Saturn's rings in amazing detail. They also discovered that there were active volcanoes on Io, one of Jupiter's moons.

Where Are They Now?

The *Voyager* probes are still exploring the universe. *Voyager 2* flew past Uranus and Neptune. It is now in a part of space called the heliosphere, right at the edge of our solar system. *Voyager 1* is now in **interstellar** space, which is outside our solar system—around 12 billion miles (19 billion kilometers) from our Sun. Both probes are still sending information home.

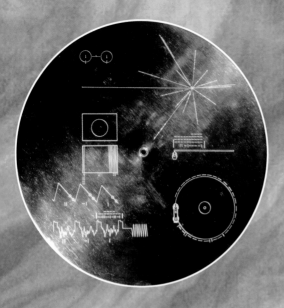

◄ *The Voyager probes carry disks with pictures and sounds from Earth on them, in case they come across other life-forms.*

NASA's *New Horizons* space probe set off for Pluto in January 2006. It finally reached its destination in 2015. It spent six months studying the dwarf planet, and is now heading toward the very edge of the solar system.

interstellar: among the stars, beyond our solar system

Space Planes

Until the 1980s, spacecraft could only be used for one mission. In 1983, NASA revealed the first reusable space vehicle—the Space Shuttle. They built a **fleet** of six shuttles that were designed to carry astronauts and supplies to and from Earth's orbit. Each shuttle could be used around 100 times. Powered by five massive rocket motors, they could travel at 17,000 mph (27,500 km/h).

◀ *The Space Shuttles looked more like airplanes than rockets.*

Shuttle Disasters

In January 1986, the shuttle *Challenger* was setting off on a routine mission. Just after launch, the fuel system failed. The shuttle exploded, killing all seven astronauts on board. Disaster struck again in 2003. The shuttle *Columbia* broke apart as it was coming back from a mission. Seven astronauts also died in this tragedy.

"They had a hunger to explore the universe and discover its truths. They wished to serve, and they did."

President Ronald Reagan, talking about the *Challenger* astronauts

▲ *This is a poster created in memory of the crew of the Space Shuttle* Challenger.

fleet: a group of vehicles

▲ *Astronauts often stay on space stations for months at a time.*

Space Stations

A space station is a special kind of spacecraft. It is not
designed for traveling into deep space. It is meant to stay in
orbit around Earth. Scientists use space stations as research
centers. The stations have **laboratories** and all the things
astronauts need to live in space.

The ISS

Construction of the International Space Station (ISS) began
in 1986. It was built in space, piece by piece. The first crew
moved into the ISS in 2000, but the last piece was not added
until 2010. Astronauts from many different countries work
together in the ISS to understand more about space.

The USSR also launched a space station in 1986. It was called Mir. This name means both "peace" and "world" in Russian. Mir stayed in orbit for 15 years, before crashing to Earth in 2001.

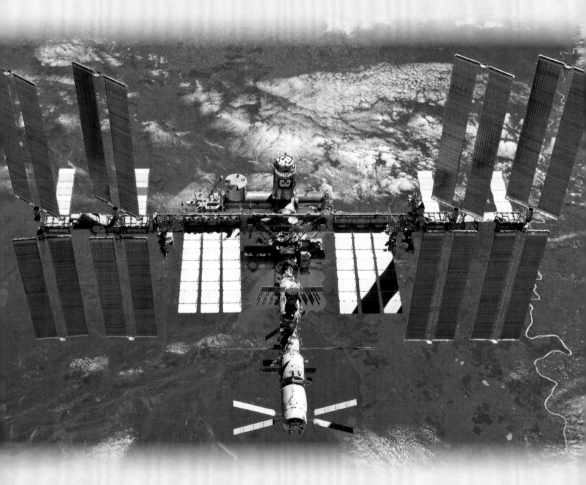

▲ *The ISS travels around Earth at a speed of 17,000 mph (27,400 km/h).*

laboratories: rooms where scientists carry out experiments

Mars and Beyond

To Boldly Go...

We have put men on the Moon. We have probed the outer planets. We have even sent spacecraft beyond our solar system. As technology continues to develop and our understanding of space grows, what might we discover next? How far will we travel?

▶ *Hubble has captured extraordinary images of the universe, such as these remains of a star that exploded.*

Hubble is a giant telescope orbiting Earth. It can see deeper into space than any other telescope. Hubble has taken incredible pictures of the creation of stars and of faraway **galaxies**.

Space Agencies

The USA and Russia (the former USSR) are no longer the only countries with space agencies like NASA. The European Space Agency, as well as countries such as China, India, and Israel, all have space programs. Private companies are also starting to fund space exploration.

▲ *Today, Russian and European astronauts work alongside NASA astronauts on the ISS.*

galaxies: huge groups of stars, gas, and dust

Early Mars Missions

Is it possible to live on another planet? Scientists think it might be. Since the 1970s, spacecraft have been exploring Mars to see if it might support human life. The very first Mars mission was the Soviet probe *Mars 2*. It was built in two parts—one part to land on the planet and the other to orbit it.

In 2015, a probe orbiting Mars sent back some photographs that amazed the world. The pictures suggested that water might still flow on the planet during Mars's summer.

▼ *The dark streaks on this picture of Mars may be proof that water flows there.*

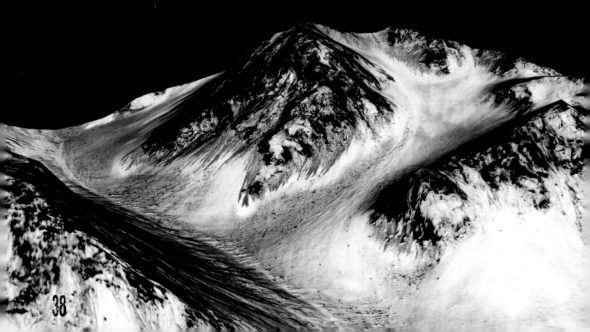

▲ *Scientists can control the Mars rovers from Earth.*

Roving Mars

In 2003, NASA landed two "rovers" on Mars. These **robotic** vehicles are called *Spirit* and *Opportunity*. They move around the surface of the planet, sending back information about the soil and any evidence of liquid water. Liquid water is the key to life. Until 2015, scientists did not think it existed on Mars, but they were hoping for evidence that water had existed there in the past.

robotic: describing a machine that is programmed to do certain tasks

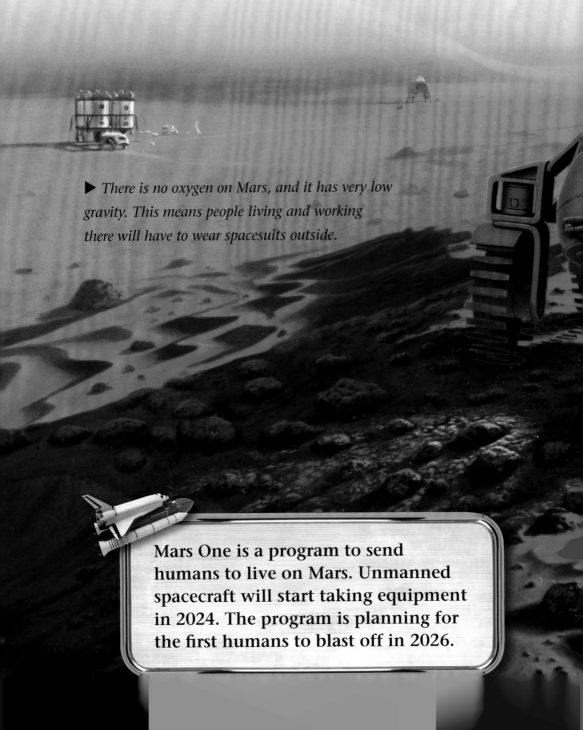

What's the Problem?

No one has ever walked on Mars, but that could soon change. NASA is preparing for humans to explore, and later live on, the planet. They are finding ways to protect astronauts from the **radiation** on Mars, but there are still many other challenges that humans will face.

▶ *There is no oxygen on Mars, and it has very low gravity. This means people living and working there will have to wear spacesuits outside.*

Mars One is a program to send humans to live on Mars. Unmanned spacecraft will start taking equipment in 2024. The program is planning for the first humans to blast off in 2026.

Life on Mars

Exploring Mars will be a tough mission. Astronauts will be away for years, not days like they were for the Moon missions. They will have to live in crowded conditions with the same small group of people. Astronauts on a mission like this will have to go through many tests to make sure they are up to the job!

radiation: a type of energy that can be harmful to people

Europa Mission

Mars might not be the only other place in space that people could live. In the next 10 years, NASA hopes to launch a mission to Jupiter. The plan is to find out more about the giant planet's moon, Europa. There might be liquid water beneath Europa's icy surface.

▶ **Asteroids** *are also known as "minor planets."*

The European Space Agency sent a probe called Rosetta to Comet 67P/Churyumov–Gerasimenko. After 10 years, it arrived in 2014. The probe sent a small device to the surface of the comet, which sent back information to Earth.

Landing on an Asteroid

NASA is also hoping to visit an asteroid by 2025. First, an unmanned mission will break off a large boulder from an **asteroid**. It will then put it into orbit around the Moon. Once there, humans will be able to study it more easily. A special spacecraft called *Orion* is being built to carry astronauts to the asteroid. NASA hopes that Orion will also be able to carry astronauts deeper into space than ever before.

▶ *If there is water on the moon Europa, humans may one day be able to live there.*

asteroids: rocky bodies, smaller than a planet, orbiting the Sun

Space Tourism

Space agencies such as NASA will continue to explore deep space, but private companies are now starting to fund near-Earth space transportation. This includes space tourism. In 2001, American millionaire, Dennis Tito, became the world's first space tourist. He **trained** for 900 hours and paid $20 million to spend six days on the International Space Station.

In October 2014, during a test flight, one of Virgin Galactic's SpaceShipTwo "spaceplanes" broke up while in the air. It crashed in the desert, killing one of its pilots.

◄ *Virgin Galactic is planning to build a fleet of SpaceShipTwo craft to carry people into space.*

Going Galactic

Soon, other people will have the chance to tour space. Some private companies, like Virgin Galactic, are accepting applications for future astronauts—if they have $250,000 to spare! Perhaps within the next few decades, traveling beyond Earth will be as common as flying by airplane.

trained: practiced doing something

Learning More

Books

Discovering Mars
by Melvin Berger and Mary Kay Carson
(Scholastic, 2015)

Everything Space
by Helaine Becker
(National Geographic, 2015)

The First Moon Walk
by Ryan Nagelhout
(Gareth Stevens, 2015)

Websites

www.cosmos4kids.com/files/explore_intro.html
All about space, including space exploration, our solar system, and beyond.

http://mars.nasa.gov/participate/funzone/
Learn about the red planet on NASA's Mars Exploration pages.

http://solarsystem.nasa.gov/kids/
Take a journey through our solar system on NASA's kids pages.

Glossary

asteroids Rocky bodies, smaller than a planet, orbiting the Sun

astronomers Scientists who study objects in space

CapCom: Capsule Communicator—the person on Earth in charge of talking to astronauts in space

command module The control center and living quarters in a spacecraft

cosmonaut A Russian astronaut

eclipses When an object in space blocks the light from another object

EVA When astronauts move around in space outside their spacecraft

fleet A group of vehicles

galaxies Huge groups of stars, gas, and dust

gravity The force of attraction between two objects

horizon The line where land and sky seem to meet

interstellar Among the stars, beyond our solar system

laboratories Rooms where scientists carry out experiments

orbit To move around a circular object

probe A small, unmanned spacecraft

radiation A type of energy that can be harmful to people

robotic Describing a machine that can be programmed to do certain tasks

satellite A device that moves around Earth, collecting or sending information

splashdown When a spacecraft lands in the sea

trained Practiced doing something

transmissions Radio messages

Index

Entries in **bold** refer to pictures